ENERGY SECTOR STANDARD
OF THE PEOPLE'S REPUBLIC OF CHINA

中华人民共和国能源行业标准

Technical Code for Soil and Water Conservation of Hydropower Construction Projects

水电建设项目水土保持技术规范

NB/T 10509-2021

Replace DL/T 5419-2009

Chief Development Department: China Renewable Energy Engineering Institute

Approval Department: National Energy Administration of the People's Republic of China

Implementation Date: July 1, 2021

China Water & Power Press

中国水利水电出版社

Beijing 2024

All rights reserved. No part of this publication may be reproduced, stored in a retrieval system, or transmitted in any form or by any means—electronic, mechanical, photocopying, recording or otherwise, without prior written permission of the publisher.

图书在版编目（CIP）数据

水电建设项目水土保持技术规范：NB/T 10509-2021 代替 DL/T 5419-2009 = Technical Code for Soil and Water Conservation of Hydropower Construction Projects (NB/T 10509-2021 Replace DL/T 5419-2009)：英文 / 国家能源局发布. -- 北京：中国水利水电出版社, 2024. 7. -- ISBN 978-7-5226-2646-8

Ⅰ. TV-65；S157-65

中国国家版本馆CIP数据核字第2024SS7219号

ENERGY SECTOR STANDARD
OF THE PEOPLE'S REPUBLIC OF CHINA
中华人民共和国能源行业标准

Technical Code for Soil and Water Conservation
of Hydropower Construction Projects
水电建设项目水土保持技术规范
NB/T 10509-2021
Replace DL/T 5419-2009
（英文版）

Issued by National Energy Administration of the People's Republic of China
国家能源局　发布
Translation organized by China Renewable Energy Engineering Institute
水电水利规划设计总院　组织翻译
Published by China Water & Power Press
中国水利水电出版社　出版发行
　　Tel: (+ 86 10) 68545888　68545874
　　sales@mwr.gov.cn
　　Account name: China Water & Power Press
　　Address: No.1, Yuyuantan Nanlu, Haidian District, Beijing 100038, China
　　http://www.waterpub.com.cn
中国水利水电出版社微机排版中心　排版
北京中献拓方科技发展有限公司　印刷
184mm×260mm　16 开本　5.25 印张　166 千字
2024 年 7 月第 1 版　2024 年 7 月第 1 次印刷
Price（定价）：￥860.00

Introduction

This English version is one of China's energy sector standard series in English. Its translation was organized by China Renewable Energy Engineering Institute authorized by National Energy Administration of the People's Republic of China in compliance with relevant procedures and stipulations. This version was issued by National Energy Administration of the People's Republic of China in Announcement [2023] No. 1 dated February 6, 2023.

This version was translated from the Chinese standard NB/T 10509-2021, *Technical Code for Soil and Water Conservation of Hydropower Construction Projects*, published by China Water & Power Press. The copyright is reserved by National Energy Administration of the People's Republic of China. In the event of any discrepancy in the implementation, the Chinese version shall prevail.

Many thanks go to the staff from the relevant standard development organizations and those who have provided generous assistance in the translation and review process.

For further improvement of the English version, any comments and suggestions are welcome and should be addressed to:

China Renewable Energy Engineering Institute
No. 2 Beixiaojie, Liupukang, Xicheng District, Beijing 100120, China
Website: www.creei.cn

Translating organizations:

POWERCHINA Huadong Engineering Corporation Limited

China Renewable Energy Engineering Institute

Translating staff:

ZENG Yang	CHEN Ni	SONG Xiaoyan	YANG Kai
YING Feng	LI Jian	SHI Lingtong	ZHAO Man
GUO Zihao	LUO Yiwei	PENG Le	SHAN Jie

Review panel members:

LIU Xiaofen	POWERCHINA Zhongnan Engineering Corporation Limited
CUI Lei	China Renewable Energy Engineering Institute
YAN Wenjun	Army Academy of Armored Forces, PLA

JIA Haibo	POWERCHINA Kunming Engineering Corporation Limited
SHE Dongli	Hohai University
GAO Yan	POWERCHINA Beijing Engineering Corporation Limited
LI Qian	POWERCHINA Chengdu Engineering Corporation Limited
LI Shisheng	China Renewable Energy Engineering Institute

National Energy Administration of the People's Republic of China

翻译出版说明

本译本为国家能源局委托水电水利规划设计总院按照有关程序和规定，统一组织翻译的能源行业标准英文版系列译本之一。2023年2月6日，国家能源局以2023年第1号公告予以公布。

本译本是根据中国水利水电出版社出版的《水电建设项目水土保持技术规范》NB/T 10509—2021翻译的，著作权归国家能源局所有。在使用过程中，如出现异议，以中文版为准。

本译本在翻译和审核过程中，本标准编制单位及编制组有关成员给予了积极协助。

为不断提高本译本的质量，欢迎使用者提出意见和建议，并反馈给水电水利规划设计总院。

地址：北京市西城区六铺炕北小街2号
邮编：100120
网址：www.creei.cn

本译本翻译单位：中国电建集团华东勘测设计研究院有限公司
水电水利规划设计总院

本译本翻译人员：曾旸　陈妮　宋晓彦　杨　凯
应　丰　李　健　侍灵通　赵　满
郭子豪　罗艺伟　彭　玏　单　婕

本译本审核人员：

刘小芬　中国电建集团中南勘测设计研究院有限公司
崔　磊　水电水利规划设计总院
闫文军　中国人民解放军陆军装甲兵学院
贾海波　中国电建集团昆明勘测设计研究院有限公司
佘冬立　河海大学
高　燕　中国电建集团北京勘测设计研究院有限公司
李　茜　中国电建集团成都勘测设计研究院有限公司
李仕胜　水电水利规划设计总院

国家能源局

Announcement of National Energy Administration of the People's Republic of China [2021] No. 1

National Energy Administration of the People's Republic of China has approved and issued 320 energy sector standards including *Code for Integrated Resettlement Design of Hydropower Projects* (Attachment 1), the foreign language versions of 113 energy sector standards including *Carbon Steel and Low Alloy Steel for Pressurized Water Reactor Nuclear Power Plants—Part 7: Class 1, 2, 3 Plates* (Attachment 2), and the amendment notification for 5 energy sector standards including *Technical Code for Investigation and Assessment of Aquatic Ecosystem for Hydropower Projects* (Attachment 3).

Attachments: 1. Directory of Sector Standards
              2. Directory of Foreign Language Versions of Sector Standards
              3. Amendment Notification of Sector Standards

National Energy Administration of the People's Republic of China

January 7, 2021

Attachment 1:

Directory of Sector Standards

Serial number	Standard No.	Title	Replaced standard No.	Adopted international standard No.	Approval date	Implementation date
...						
26	NB/T 10509-2021	Technical Code for Soil and Water Conservation of Hydropower Construction Projects	DL/T 5419-2009		2021-01-07	2021-07-01
...						

Foreword

According to the requirements of Document GNKJ [2015] No. 283 issued by National Energy Administration of the People's Republic of China, "Notice on Releasing the Development and Revision Plan of Energy Sector Standards in 2015", and after extensive investigation and research, summarization of practical experience, and wide solicitation of opinions, the drafting group has prepared this code.

The main technical contents in this code include: general provisions, terms, basic requirements, classification and design standards of soil and water conservation works, assessment on soil and water conservation, range of responsibility for soil and water conservation and zoning for soil erosion control, prediction and analysis of soil erosion and water loss, goals and measures layout of soil erosion and water control, residues disposal area, protection and utilization of surface soil resources, design of soil erosion and water loss control measures, construction planning on soil and water conservation, soil and water conservation supervision, soil and water conservation monitoring, scientific research on soil and water conservation, general design of soil and water conservation, cost and benefit analysis of soil and water conservation, and management of soil and water conservation works.

The main technical contents revised are as follows:

—Adding seven chapters: classification and design standards of soil and water conservation works, residues disposal area, protection and utilization of surface soil resources, soil and water conservation supervision, scientific research on soil and water conservation, general design of soil and water conservation, and management of soil and water conservation works.

—Adjusting the application scope to "soil and water conservation of hydropower construction projects".

National Energy Administration of the People's Republic of China is in charge of the administration of this code. China Renewable Energy Engineering Institute has proposed this code and is responsible for its routine management. Energy Sector Standardization Technical Committee on Hydropower Planning, Resettlement and Environmental Protection (NEA/TC16) is responsible for the explanation of specific technical contents. Comments and suggestions in the implementation of this code should be addressed to:

China Renewable Energy Engineering Institute
No. 2 Beixiaojie, Liupukang, Xicheng District, Beijing 100120, China

Chief development organizations:

    POWERCHINA Huadong Engineering Corporation Limited

    China Renewable Energy Engineering Institute

Participating development organizations:

    POWERCHINA Northwest Engineering Corporation Limited

    POWERCHINA Chengdu Engineering Corporation Limited

    POWERCHINA Guiyang Engineering Corporation Limited

    POWERCHINA Zhongnan Engineering Corporation Limited

    POWERCHINA Kunming Engineering Corporation Limited

    POWERCHINA Beijing Engineering Corporation Limited

Chief drafting staff:

YING Feng	Cui Lei	LI Jian	WANG Jing
Qian Aiguo	YANG Kai	DU Yunling	LI Hailin
NIU Zhenhua	XIONG Ying	SHAN Jie	LIAO Qichen
WANG Zheng	ZHAO Jun	FENG Lei	GAO Fan
CHEN Pingping	ZOU Binghua	LI Hongxing	WANG Yuyan
YUE Zengbi	JIA Yanghai	GENG Xiangguo	ZHAO Xiaohong
YAN Qiao	LIU Jian	TAO Chunwei	YU Qian
ZHANG Yi	QIN Yang	HAN Xiaojie	ZHANG Jun
MA Shijun	ZHOU Yongfeng	SHAO Yunzhe	HAO Yuejiao
CHEN Ni	LI Caixia	FENG Jingen	LI Xin
YAN Yan	XIA Zhaohui	WU Derong	ZHANG Yu
CHEN Dong	WU Wei	FEI Ripeng	ZENG Yang

Review panel members:

WAN Wengong	JIN Yi	CHEN Shengli	SHI Jiayue
YAN Junping	ZHANG Naichang	CHEN Fan	ZHANG Hucheng
XIONG Heyong	ZHANG Xiaoli	ZHU Yonggang	NIU Junwen
HE Kangning	ZHENG Cheng	HAO Lian'an	REN Lyu
ZHOU Xiangshan	LI Shisheng		

Contents

1	**General Provisions**	1
2	**Terms**	2
3	**Basic Requirements**	3
3.1	General Requirements	3
3.2	Basic Data Investigation and Survey	4
3.3	Design Depth and Requirements of Each Stage	7
3.4	Soil and Water Conservation Program	13
3.5	Soil and Water Conservation of Resettlement Area	14
4	**Classification and Design Standards of Soil and Water Conservation Works**	**16**
4.1	Classification of Soil and Water Conservation Works	16
4.2	Design Standards	16
4.3	Hydrological Calculation	16
5	**Assessment on Soil and Water Conservation**	**18**
5.1	General Requirements	18
5.2	Assessment Requirements of Each Design Stage	19
5.3	Characteristics and Key Points of Analysis and Assessment for Hydropower Construction Project in Special Areas	19
6	**Range of Responsibility for Soil and Water Conservation and Zoning of Soil Erosion Control**	**21**
6.1	General Requirements	21
6.2	Identification of Range of Responsibility for Soil and Water Conservation	21
6.3	Zoning of Soil Erosion Control	21
7	**Prediction and Analysis of Soil Erosion and Water Loss**	**23**
7.1	General Requirements	23
7.2	Basic Requirements of Each Stage	23
7.3	Analysis of Prediction Results	24
8	**Goals and Measures Layout of Soil Erosion and Water Loss Control**	**26**
8.1	General Requirements	26
8.2	Goals and Standards of Soil Erosion and Water Loss Control	26
8.3	Layout of Soil Erosion and Water Loss Control Measures	26
9	**Residues Disposal Area**	**30**
10	**Protection and Utilization of Surface Soil Resources**	**32**
11	**Design of Soil Erosion and Water Loss Control Measures**	**33**
11.1	Dregs-Blocking Works	33

11.2	Flood Release and Diversion Works	33
11.3	Slope Protection Works	34
11.4	Precipitation Storage and Infiltration Works	34
11.5	Land Improvement Works	34
11.6	Windbreak and Sand Fixation Works	35
11.7	Vegetation Construction Works	35
11.8	Temporary Protection Works	36
12	**Construction Planning on Soil and Water Conservation**	**37**
12.1	General Requirements	37
12.2	Technical Requirements	37
12.3	Implementation Requirements	39
13	**Soil and Water Conservation Supervision**	**40**
13.1	General Requirements	40
13.2	Basic Requirements of Each Stage	40
13.3	Supervision Implementation	41
14	**Soil and Water Conservation Monitoring**	**42**
15	**Scientific Research on Soil and Water Conservation**	**43**
15.1	General Requirements	43
15.2	Scientific Research Requirements and Tasks of Each Stage	43
15.3	Scientific Research Implementation	44
16	**General Design of Soil and Water Conservation**	**45**
16.1	General Requirements	45
16.2	Technical Requirements for General Design of Soil and Water Conservation	45
16.3	Technical Requirements for "Three-Simultaneity" Implementation Plan of Soil and Water Conservation	46
17	**Cost and Benefit Analysis of Soil and Water Conservation**	**48**
17.1	General Requirements	48
17.2	Cost of Soil and Water Conservation	48
17.3	Benefit Analysis of Soil and Water Conservation	48
18	**Management of Soil and Water Conservation Works**	**50**
18.1	General Requirements	50
18.2	Management in Construction Period	50
18.3	Management in Operation Period	52
Appendix A	**Contents of Soil and Water Conservation Chapter in the Hydropower Planning Report**	**54**
Appendix B	**Contents of Soil and Water Conservation Chapter**	

		in the Pre-feasibility Study Report of Hydropower Construction Project ·················· 55
Appendix C		Contents of Soil and Water Conservation Chapter in the Feasibility Study Report of Hydropower Construction Project ·················· 56
Appendix D		Contents of General Design and "Three-Simultaneity" Implementation Plan on Soil and Water Conservation of Hydropower Construction Project ·················· 58
Appendix E		Contents of Explanation of Construction Drawing Design of Soil and Water Conservation Works of Hydropower Construction Project ·············· 61
Appendix F		Contents of Summary Report on Soil and Water Conservation Design of Hydropower Construction Project ·················· 62
Appendix G		Contents of Post Assessment Report on Soil and Water Conservation of Hydropower Construction Project ·················· 63
Appendix H		Contents of Soil and Water Conservation Program of Hydropower Construction Project ············· 64

Explanation of Wording in This Code·················· 67
List of Quoted Standards ·················· 68

1 General Provisions

1.0.1 This code is formulated with a view to standardizing the soil and water conservation, and unifying the content, methods and technical requirements of soil and water conservation for hydropower construction projects, so as to prevent and control soil erosion caused by the construction activities of hydropower construction projects, prevent and control the hazards of soil erosion, and restore and improve the ecological environment of the project area.

1.0.2 This code is applicable to the soil and water conservation for hydropower construction projects.

1.0.3 Soil and water conservation for hydropower construction projects shall include the design of soil and water conservation works and the preparation of soil and water conservation program at the stages of planning, pre-feasibility study, feasibility study, detailed design; and soil and water conservation in the whole process of construction, acceptance, operation and decommissioning of the project.

1.0.4 Soil and water conservation for hydropower construction projects shall adhere to the principles of prevention, comprehensive control, safety and reliability, ecological priority, adaptation to local conditions, technological feasibility and economic rationality, formulate the objectives of soil erosion control, and establish and implement the system of soil and water conservation measures.

1.0.5 Soil erosion control for hydropower construction projects shall highlight investigation and study, and encourage the adoption of new technologies, new processes, new equipment and new materials.

1.0.6 In addition to this code, the soil and water conservation of hydropower construction projects shall comply with other current relevant standards of China.

2 Terms

2.0.1 post assessment on soil and water conservation

assessment on the construction, control effectiveness, objective achievements, benefits and sustainability of soil and water conservation facilities, which have passed the acceptance and been put into use, by scientific, systematic and standardized methods and means

3 Basic Requirements

3.1 General Requirements

3.1.1 The technical work of soil and water conservation of hydropower construction projects shall mainly include the program preparation, design, construction, supervision, monitoring, and acceptance of soil and water conservation works. For a hydropower construction project, the soil and water conservation works shall be carried out synchronously with the main works at each stage, and highlight the whole process control and informatization management.

3.1.2 The alternatives comparison and design of main works shall consider the influence and constraints of soil erosion and meet the requirements of soil and water conservation.

3.1.3 The soil erosion control for hydropower construction projects shall meet the following requirements:

1. Disturbance and damage to the original landform and surface vegetation shall be controlled and reduced, occupation of soil and water resources shall be reduced, and improvement of the resource utilization efficiency shall be emphasized.

2. For the ecologically fragile areas where the original surface vegetation is difficult to restore after being damaged, transplanting and protection measures shall be taken for the disturbed vegetation according to the actual situation of the project area, so as to facilitate the later restoration and utilization of vegetation.

3. Priority shall be given to the comprehensive utilization of earth-rock excavated from the project area to reduce residues. Sites shall be designated to stockpile and dispose of residues. Blocking measures and flood control and drainage measures shall be taken following the principle of "blocking before abandoning".

4. Ecological restoration measures shall be taken for slope protection on the premise of satisfying the suitable growth conditions, meeting the main functional requirements and not affecting the safety of the project.

5. Protective measures such as blocking, slope protection, interception and drainage shall be adopted for the excavation, filling, stockpiling and disposal sites.

6. The temporary measures of blocking, drainage, sediment deposition,

covering, and vegetation shall be taken in the construction period.

7 Upon the completion of project construction, the construction sites shall be reclaimed or restored according to the original land use function or the planned land use.

3.1.4 The construction planning of main works shall meet the following requirements:

1 Where a diversion cofferdam is set, its filling type, earth-rock source, protection plan, demolition method, and residues disposal shall be clearly defined.

2 Quarrying and borrow area excavation shall adopt benched mining according to the engineering geology, and create suitable growth conditions and construction conditions for soil erosion control and vegetation restoration later.

3 The land occupation of temporary roads and production and living facilities shall be strictly controlled, so as to reduce the disturbance to the original landform and the destruction of surface vegetation, and temporary protective measures shall be taken.

4 Temporary stockpile sites shall be set for the storage of earth-rock that needs rehandling, comprehensive utilization or processing utilization, for which the construction layout shall be defined and temporary protection shall be provided.

5 The construction procedure and methods resulting in less disturbance to the ground surface and less residues shall be adopted.

6 The earth-rock construction in the area prone to water erosion should avoid high rainfall seasons; earth-rock construction in the area prone to wind erosion should avoid gale seasons; the time of the surface exposure in the above seasons shall be reduced, and temporary protection shall be strengthened in case of rainstorm or strong wind.

3.1.5 The design of soil and water conservation for hydropower construction projects shall comply with the current sector standard NB/T 10344, *Code for Design of Soil and Water Conservation for Hydropower Projects*.

3.2 Basic Data Investigation and Survey

3.2.1 Basic data investigation and survey shall comply with the current relevant investigation and survey standards of the nation, hydropower sector and soil and water conservation sector.

3.2.2 The basic data shall be obtained according to the following requirements:

- 1 Basic data shall be obtained by means of collection, on-site investigation and survey, and adopted after analysis and demonstration.

- 2 The data on the survey of main works, hydrology and sediment, engineering planning, hydraulic structure design, construction planning, resettlement, environmental protection, project management, and project cost shall be collected.

- 3 The data on the natural environment, social economy, soil erosion, current situation of soil and water conservation, sensitive areas of soil and water conservation, ecologically fragile areas, and important ecological function areas shall be collected.

- 4 The investigation of sensitive areas of soil and water conservation, ecologically fragile areas and important ecological function areas mainly includes the following:

    1) The investigation of sensitive areas of soil and water conservation shall ascertain whether the project area involves key prevention and control areas of soil erosion, drinking water source protection area, protected and reserved areas in first-grade water function zones, nature reserves, world cultural and natural heritage sites, scenic spots, geological park, forest park, glacier park, important wetlands, ecological red lines, etc.

    2) The investigation of ecologically fragile areas shall ascertain whether the project area belongs to the ecologically fragile areas such as the forest and grass interaction zones in northeast China, the farming and pastoral interaction zones in north China, the desert and oasis interaction zones in northwest China, the rock desertification zones in the karst mountain in southwest China, the farming and pastoral interaction zones in mountainous areas of southwest China, and the compound erosion zones of the Qinghai-Xizang Plateau in China.

    3) The investigation of important ecological function areas shall ascertain whether the project area involves important ecological regulation functions such as water source conservation, biodiversity conservation, soil conservation, windbreak and sand fixation, and flood regulation and storage.

3.2.3 The survey at the pre-feasibility study stage shall meet the following requirements:

- 1 Topographic maps of residues disposal areas, quarries, borrow areas, and other key soil erosion areas shall be collected and the scale shall not be less than 1 : 10 000. The scope shall cover important protected objects such as surrounding settlements, farmlands, roads, rivers, etc.

- 2 Geological data of residues disposal areas shall be collected and, if necessary, the level of investigation shall meet the requirements of the pre-feasibility study of the hydropower construction project.

3.2.4 The survey at the feasibility study stage shall meet the following requirements:

- 1 Engineering geological data and topographic maps of residues disposal areas, quarries, borrow areas, and other key soil erosion areas shall be collected. The scale of the topographic maps shall not be less than 1 : 2 000, and the scope shall cover no less than 50 m beyond the design property boundary, including surrounding settlements, farmlands, roads, rivers, etc.

- 2 Detailed survey shall be carried out for the important residues disposal areas with a residues volume of more than 500000 m^3, a stockpile height of more than 20 m, a catchment area of more than 1 km^2, or important protected objects in the downstream. The survey shall comply with the current national standard GB/T 51297, *Standard for Investigation and Survey of Soil and Water Conservation Projects*; and GB 50021, *Code for Investigation of Geotechnical Engineering*. The survey for structures shall comply with the current national standard GB 50487, *Code for Engineering Geological Investigation of Water Resources and Hydropower*. For other residues disposal areas, necessary survey shall be carried out.

3.2.5 The survey at the detailed design stage shall meet the following requirements:

- 1 The scale of topographic maps of residues disposal areas, quarries, borrow areas and other key soil erosion areas shall not be less than 1 : 1 000.

- 2 Detailed survey shall be carried out for Grades 1 to 3 residues disposal areas of hydropower construction projects, which shall comply with the current national standard GB 50021, *Code for Investigation of Geotechnical Engineering*. Necessary survey shall be carried out for Grades 4 and 5 residues disposal areas.

3   The scale of topographic maps for the design of individual measures, such as dregs blocking works and slope protection works, shall not be less than 1 : 500, and detailed geological survey drawings and data shall be provided.

3.2.6   At the feasibility study stage, the soil and water conservation works investigation and geological survey report, the surface soil resources investigation report, and the plant resources investigation report shall be provided as the basic data for the soil and water conservation design at the feasibility study stage.

3.2.7   Investigation and survey of soil and water conservation for hydropower construction projects shall be undertaken by units with relevant qualifications or competence.

3.3   Design Depth and Requirements of Each Stage

3.3.1   The soil and water conservation work at the planning stage shall meet the following requirements:

1   The work content shall mainly include the investigation of the sensitivity and constraints of soil and water conservation, analysis of the effects and constraints of soil erosion, putting forward the preliminary countermeasures, as well as relevant planning for scientific research and special studies for the project involving sensitive areas of soil and water conservation, ecologically fragile areas, important eco-functional areas, and with constraints on soil and water conservation.

2   The key points of the work shall mainly include the influence of the project construction on soil erosion, the investigation of the constraints on soil and water conservation, and the guidelines and countermeasures for soil erosion control.

3   The outcome is the soil and water conservation chapter of the hydropower planning report. The contents of soil and water conservation chapter in the hydropower planning report shall be in accordance with Appendix A of this code.

3.3.2   The soil and water conservation work at the pre-feasibility study stage shall meet the following requirements:

1   The work content shall mainly include the investigation and analysis of the natural conditions and soil erosion factors, analysis of soil and water conservation, preliminarily defining the responsibility range and zoning of soil erosion control, preliminarily proposing the grades

and objectives for soil erosion control, formulating the soil and water conservation measures by zones, formulating the monitoring and management plan for soil and water conservation, estimation of soil and water conservation investment, putting forward preliminary conclusions of soil and water conservation and pending issues in the following stages and the suggestions to deal with them, putting forward requirements and suggestions on the follow-up survey, design and scientific research planning.

2   The key points of the work shall mainly include putting forward in-depth solutions and measures for the problems in the planning stage and the soil erosion problems existing in the project, preliminary investigation of the surface soil resources and estimation of the amount of surface soil stripping, and preliminary determination of the sites of residues disposal areas and surface soil storage areas. Specific scientific research subjects and special study plans shall be put forward for the project involving sensitive areas of soil and water conservation, ecologically fragile areas, important ecological function areas, and constraints on soil and water conservation, and relevant work shall be initiated timely. Interim outcomes shall be provided for major scientific research and special demonstration.

3   The outcomes shall mainly include the interim outcomes of major research and special demonstration on soil and water conservation, and the soil and water conservation chapter in the pre-feasibility study report. The contents of soil and water conservation chapter in the pre-feasibility study report of hydropower construction project shall be in accordance with Appendix B of this code.

3.3.3   The soil and water conservation work at the feasibility study stage shall meet the following requirements:

1   The work content shall mainly include carrying out the analysis and evaluation of soil and water conservation, defining the responsibility range and key points for soil erosion control, determining the grades and objectives for soil erosion control, determining the system and layout of soil and water conservation measures, formulating soil and water conservation measures by zones, determining the grade, design standard and structural type of soil and water conservation works, conducting the construction planning of soil and water conservation works, determining the construction schedule of soil and water conservation works, determining the monitoring program of soil

and water conservation works, determining the work quantity and investment budget, putting forward the management opinions on the implementation of soil and water conservation works, analyzing the benefits of soil and water conservation works, and putting forward the conclusions on soil and water conservation and suggestions for follow-up work.

2 The key points of the work shall mainly include putting forward specific guidelines and countermeasures for the problems in the pre-feasible study stage, determining the design standard and works grade, investigating the surface soil resources in detail and calculating the amounts of stripping and utilization, determining the sites of residues disposal areas and surface soil storage areas with detailed design, carrying out detailed design of soil and water conservation measures by zones, and determining the cost estimate of soil and water conservation.

3 The outcomes shall mainly include the soil and water conservation chapters in the feasibility study report and relevant special reports, soil and water conservation program, soil and water conservation scientific research, and special demonstration report on soil and water conservation.

4 The soil and water conservation chapters shall be prepared for the special reports on comparison of dam site and dam type alternatives, selection of normal storage level, and planning of construction general layout at the feasibility study stage.

5 The soil and water conservation program shall be prepared at the feasibility study stage.

6 The scientific research and special demonstration on soil and water conservation at the feasibility study stage shall meet the following requirements:

1) Scientific research on soil and water conservation shall consider the needs of the hydropower construction project, and covers the researches on influencing factors and mechanism of soil erosion, soil and water conservation and ecological restoration technology for high and steep slopes, reservoir shores and ecologically connected areas, virginal vegetation protection technology, soil amelioration technology, etc.

2) The special demonstration of soil and water conservation shall be carried out considering the needs of the hydropower construction

project, including the special studies on the protection design of residues disposal areas, the planning of surface soil resource stripping and protection, etc.

3) The outcomes of scientific research and special demonstration on soil and water conservation shall be incorporated into the soil and water conservation program to support the design of soil and water conservation and the implementation of soil and water conservation measures.

7 The design outcomes shall mainly include the soil and water conservation chapter in the feasibility study report of the hydropower construction project, and the outcomes of scientific research and special demonstration of soil and water conservation. The contents of soil and water conservation chapter in the feasibility study report of hydropower construction project shall be in accordance with Appendix C of this code.

3.3.4 The work of soil and water conservation at the stages of detailed design and construction shall meet the following requirements:

1 The requirements of the approved soil and water conservation program shall be fully implemented, the design of construction drawings related to soil and water conservation shall be completed, and the outcomes of relevant scientific research and special demonstration at the planning, pre-feasibility study and feasibility study stages shall be applied.

2 The construction planning and construction of soil and water conservation works shall be done well, the supervision and monitoring of soil and water conservation as well as soil erosion control shall be implemented, and the construction management of soil and water conservation works shall be done well.

3 The general design of soil and water conservation and "three-simultaneity" implementation plan shall be prepared, and the detailed content, organization arrangement, schedule and cost of soil and water conservation shall be planned in a holistic manner. The contents of general design and "three-simultaneity" implementation plan on soil and water conservation of hydropower construction project shall be in accordance with Appendix D of this code.

4 If the soil and water conservation works is included in the bidding design document of main works, the technical terms, relevant drawings and bill of quantities of the soil and water conservation works, the

requirements for the later demolition and restoration of the construction sites, temporary roads and other temporary facilities, as well as the requirements for measurement and payment shall be implemented by bids of the main works.

5 If the soil and water conservation works constitutes a separate bid, the bidding documents shall be prepared, the main contents of which shall include general technical terms, special technical terms, bidding design drawings, bill of quantities, measurement, and payment.

6 The ecological restoration of soil and water conservation and the protection and utilization of surface soil resources shall comply with the current sector standard NB/T 10510, *Technical Code for Eco-Restoration of Soil and Water Conservation for Hydropower Projects*.

7 For the residues disposal area with a residues volume of more than 500000 m^3, a maximum stockpile height of more than 20 m, or a catchment area of more than 1 km^2, and with important protected objects downstream, the stability of the residues disposal area shall be checked, and a verification report shall be provided accordingly.

8 For special soil and water conservation works, construction drawings shall be designed and the description of construction drawing design shall be prepared. The design drawings shall mainly include the layout plan, sections, structural diagrams, details, reinforcement diagrams, and vegetation measures construction drawings of individual works. The contents of explanation of construction drawing design of soil and water conservation works of hydropower construction project shall be in accordance with Appendix E of this code.

9 The relevant design work for the special soil and water conservation works, such as the bidding documents, construction drawing design and the verification report on stability design of residues disposal areas, shall be undertaken by competent units of soil and water conservation survey and design.

3.3.5 The soil and water conservation work of hydropower construction project at the acceptance stage shall meet the following requirements:

1 The soil and water conservation work of hydropower construction project shall pass the stage acceptance and completion acceptance respectively before the stage acceptance and completion acceptance of the main works. The contents and procedures of stage acceptance and completion acceptance of soil and water conservation works shall

comply with the current sector standard NB/T 35119, *Acceptance Specification for Soil and Water Conservation Engineering of Hydropower Projects*.

2 A summary report on soil and water conservation design should be prepared before the stage acceptance and completion acceptance of soil and water conservation for the hydropower construction project. The contents of summary report on soil and water conservation design of hydropower construction project shall be in accordance with Appendix F of this code.

3.3.6 The soil and water conservation work at the operation stage of hydropower construction project shall meet the following requirements:

1 The key points of soil and water conservation at the operation stage of hydropower construction project shall mainly include:

   1) All soil and water conservation facilities built for the hydropower construction project shall be managed, inspected and maintained.

   2) Regular monitoring and evaluation of the operation and soil erosion control effects of soil and water conservation works shall be conducted for residues disposal areas with a residues volume of more than 500000 m^3, a maximum stockpile height of more than 20 m, or a catchment area of more than 1 km^2, and with important protected objects downstream, and for important soil and water conservation facilities. The annual monitoring results should be provided.

   3) Post assessment on soil and water conservation should be carried out 3 to 5 years after the special acceptance of soil and water conservation works of the hydropower construction project. The contents of post assessment report on soil and water conservation of hydropower construction projects shall be in accordance with Appendix G of this code.

2 The basis and method for post assessment on soil and water conservation for hydropower construction project shall meet the following requirements:

   1) Assessment shall be carried out by using relevant data from the stages of project planning, design, construction, acceptance and operation according to relevant national laws and regulations and technical standards.

   2) Assessment shall be carried out mainly by the methods such as

before/after project comparison, remote sensing, cumulative impact assessment, etc. through document collection, data collection and analysis, field investigation, etc.

3 Post assessment on soil and water conservation of hydropower construction project shall mainly include assessment on objectives, process and effect:

1) The assessment on the objectives of soil and water conservation shall focus on the construction objectives of soil and water conservation works, including the realization of soil and water conservation work in the project general construction objectives and the realization of the objectives of system construction on soil and water conservation.

2) The assessment on the process of soil and water conservation shall mainly include the assessment on the process of design, construction and operation.

3) The post assessment on soil and water conservation shall focus on the effect of soil and water conservation. The effects of soil and water conservation shall mainly include ecological benefits and social benefits. The assessment on effect of soil and water conservation shall focus on vegetation restoration, soil conservation, landscape upgrading, environment improvement, safety protection, socioeconomy, etc.

4 Post assessment on soil and water conservation of hydropower construction projects shall be entrusted to competent units with corresponding proficiency and ability in soil and water conservation design and consulting.

3.4 Soil and Water Conservation Program

3.4.1 The soil and water conservation program shall be prepared at the feasibility study stage of the hydropower construction project and completed the approval procedure before project construction. The contents of soil and water conservation program of hydropower construction project shall be in accordance with Appendix H of this code.

3.4.2 The main contents of the soil and water conservation program shall include the overview of project and project area, soil and water conservation assessment, soil erosion prediction, layout of soil and water conservation measures, soil and water conservation monitoring, estimation of soil and water conservation investment, and soil and water conservation management.

3.4.3 The soil and water conservation program shall specify the responsibility range and objectives for soil erosion control.

3.4.4 The design target year of soil and water conservation shall be the year of completion of the hydropower construction project or the next year, and shall be determined comprehensively according to the completion time of the hydropower construction project and the implementation schedule of soil and water conservation measures.

3.4.5 In the soil and water conservation program, the requirements for general design of soil and water conservation works and "three-simultaneity" implementation plan shall be specified, the progress on soil and water conservation research and special demonstration shall be described, and the follow-up work plan shall be put forward.

3.4.6 In the case of major change in soil and water conservation, the approved soil and water conservation program shall be supplemented or modified accordingly.

3.5 Soil and Water Conservation of Resettlement Area

3.5.1 In the resettlement planning and design, potential effects or hazards of soil erosion shall be analyzed, and soil and water conservation measures shall be taken to protect ecology and control soil erosion in resettlement areas.

3.5.2 In the soil and water conservation chapter of the river hydropower planning report, soil and water conservation measures for the resettlement areas shall be put forward, and the main types of measures shall be defined.

3.5.3 In the soil and water conservation chapter of the pre-feasibility study report of hydropower construction project, qualitative analysis shall be made on soil erosion in the resettlement area, the system of soil and water conservation measures and monitoring plan shall be preliminarily put forward, and the investment in soil and water conservation shall be estimated.

3.5.4 The soil and water conservation in resettlement areas at the feasibility study stage shall meet the following requirements:

1 The soil and water conservation chapter in the feasibility study report shall cover the resettlement areas, including the responsibility range for soil erosion control, soil erosion prediction, zoning for soil erosion control, layout of soil and water conservation measures, design of soil and water conservation measures, and monitoring plan and investment of soil and water conservation.

2 Before any disturbance from project construction, a soil and water

conservation program shall be prepared for each site of the centralized resettlements, relocated cities and towns, industrial and mining enterprises, special item relocation or reconstruction works, etc. which has an earth or rock cut-fill of 1000 m^3 or more or occupies a land area of 0.5 hm^2 or more.

3   The soil and water conservation chapters shall be prepared for the resettlement planning outline and the resettlement planning report.

3.5.5   The soil and water conservation in resettlement areas at the construction stage shall meet the following requirements:

1   The construction drawings of soil and water conservation measures shall be designed at the resettlement implementation stage, and should include the layout plan, sections, structural diagrams, details, reinforcement diagrams, vegetation measures construction drawings and temporary control measures construction drawings of individual works.

2   Soil and water conservation measures shall be implemented in resettlement works, and soil and water conservation monitoring and supervision shall be carried out.

3.5.6   The acceptance of soil and water conservation works for resettlement works shall meet the relevant requirements for the acceptance of resettlement works of the hydropower project.

4 Classification and Design Standards of Soil and Water Conservation Works

4.1 Classification of Soil and Water Conservation Works

4.1.1 The classification of soil and water conservation works shall mainly cover the residues disposal areas, structures of residues disposal area protection works, windbreak and sand fixation works, vegetation restoration and construction works, and slope protection works.

4.1.2 The classification of residues disposal areas, structures of residues disposal area protection works, windbreak and sand fixation works, and vegetation restoration and construction works shall comply with the current sector standard NB/T 10344, *Code for Design of Soil and Water Conservation for Hydropower Projects*.

4.1.3 The classification of slope protection works shall comply with the current sector standard NB/T 10512, *Code for Slope Design of Hydropower Projects*.

4.2 Design Standards

4.2.1 The flood control standards for dregs-blocking dikes, dregs-blocking dams, dregs-blocking walls, flood retaining works and flood discharge works shall be determined according to the grades of the corresponding structures.

4.2.2 The flood control standards for temporary blocking and protection works for residues disposal areas, temporary stockpile areas and surface soil storage areas, and the drainage design standards for permanent interception and drainage measures for residues disposal areas, quarries, borrow areas, production and living sites shall be determined according to the flood control and drainage requirements of corresponding objects protected.

4.2.3 The design standard for the vegetation restoration and construction works shall be determined according to its location and the nature of the occupied land.

4.2.4 The safety standard of slope protection works for sliding stability shall comply with the current sector standard NB/T 10512, *Code for Slope Design of Hydropower Projects*.

4.3 Hydrological Calculation

4.3.1 For the calculation of design flood, hydrological data shall be collected, the characteristics of the river basin or the basic situation of the catchment area shall be investigated, and the data on historical rainstorms and floods shall be used. The hydrological data on which the design flood calculation is based shall be analyzed and processed in terms of reliability, consistency and representativeness.

4.3.2 In the design of soil and water conservation works, the design flood shall be calculated according to the flood control standard, and the design flood results shall mainly include peak discharge, flood volume and flood hydrographs.

4.3.3 The hydrological calculation shall comply with the current sector standards NB/T 35046, *Code for Calculating Design Flood of Hydropower Projects*; and NB/T 35095, *Code for Hydrologic Calculation of Small Watershed of Hydropower Projects*.

5 Assessment on Soil and Water Conservation

5.1 General Requirements

5.1.1 The assessment on soil and water conservation for hydropower construction projects shall be carried out on the basis of the design of main works, the basic information on the region where the project is located, and the relevant laws, regulations, standards and plans of soil and water conservation.

5.1.2 The level of assessment of soil and water conservation shall be consistent with the level of design of hydropower construction project at each stage.

5.1.3 The assessment of soil and water conservation shall mainly cover the site selection of main works, the construction plan, the land occupation, the cut-fill balance, the layout of quarries, borrow areas, and residues disposal areas, the construction methods and procedure, and the measures with soil and water conservation functions in the design of main works, and shall meet the following requirements:

1. The assessment on site selection of main works shall ascertain whether there are constraints on soil and water conservation for the selection of dam site and axis. If any, the adjustment requirements for the site selection or design of main works shall be proposed.

2. The assessment conclusions on soil and water conservation, including normal storage level, installed capacity, dam type, hydropower complex layout and general construction layout in the project construction plan shall be made, and the optimization suggestions shall be put forward.

3. The assessment conclusions of soil and water conservation on land occupation, cut-fill balance, construction method and procedure shall be made.

4. The assessment conclusions of soil and water conservation on the site selection of quarries, borrow areas and residues disposal areas, the mining mode of quarries, borrow areas, and the residues stockpiling plan shall be made.

5. For the measures defined with soil and water conservation function in the design of main works, their locations, quantities and costs shall be tabulated by zones.

6. The issues to be studied in the next stage as well as the solutions and suggestions shall be proposed.

5.2 Assessment Requirements of Each Design Stage

5.2.1 At the planning stage, whether the general layout and site selection of the project meet the requirements of soil and water conservation shall be preliminarily analyzed, whether there are major soil and water conservation constraints shall be analyzed, and comments and suggestions for optimization, adjustment, and solution shall be put forward.

5.2.2 The soil and water conservation assessment at the pre-feasibility study stage shall meet the following requirements:

1. According to the analysis on the site selection, construction alternatives comparison and design contents of main works, whether there are soil and water conservation constraints shall be ascertained. The changes of various factors, potential impacts on soil erosion and their hazards under different schemes shall be quantitatively or qualitatively analyzed and judged. Suggestions on site selection of main works, comparison of construction layout alternatives and adjustment shall be put forward from the perspective of soil and water conservation.

2. The construction general layout and construction procedure preliminarily determined for the main works shall be assessed, especially the site selection of residues disposal areas, quarries and borrow areas shall be rationally analyzed from the perspective of soil and water conservation, and comments and suggestions shall be put forward.

5.2.3 The soil and water conservation assessment at the feasibility study stage shall meet the following requirements:

1. The assessment of site selection of main works shall focus on the construction procedure, site layout and planning rationality of residues stockpiling, quarrying and soil borrowing, surface soil storage and utilization, and put forward corresponding comments and suggestions.

2. The construction plan and layout assessment shall focus on land occupation, earth-rock volume, disturbed ground surface and damaged vegetation area, and inundation influence; impacts and constraints of different project scale schemes on soil erosion shall be analyzed; the works and measures with soil and water conservation functions in the design of main works shall be analyzed and assessed.

5.3 Characteristics and Key Points of Analysis and Assessment for Hydropower Construction Project in Special Areas

5.3.1 For hydropower construction projects involving sensitive areas of soil and water conservation, ecologically fragile areas and important ecological functional areas, the principle of minimizing ground disturbance and vegetation damage and maintaining the dominant function of soil and water conservation shall be followed; emphasis shall be placed on the analysis of the degree of irreversible vegetation damage and serious soil erosion hazards which might be caused by project construction; the requirements for dealing with constraints on soil and water conservation and comments on the project layout shall be put forward.

5.3.2 The assessment of hydropower construction projects in the mountainous and canyon regions in southwest China shall be carried out mainly from the aspects of reducing the land occupation, controlling the disturbance range, reasonably arranging the construction sites and roads, constraints such as geological disaster in the project area, and reasonable suggestions shall be put forward.

5.3.3 For the hydropower construction projects in ecologically fragile areas with compound erosion at Qinghai-Xizang plateau, the damage and restoration feasibility of surface soil and vegetation shall be analyzed, and the requirements on the construction planning, earth-rock maneuver, construction layout shall be put forward from the aspects of strictly controlling construction range, protecting surface vegetation such as alpine meadow, preventing frozen soil and freezing injury, etc. and reasonable suggestions shall be put forward.

5.3.4 For the hydropower construction projects in other regions, the assessment shall consider the regional natural conditions and project characteristics and comply with the current national standard GB 50433, *Technical Standard of Soil and Water Conservation for Production and Construction Projects*, as well as the special provisions of different types of soil erosion areas.

6 Range of Responsibility for Soil and Water Conservation and Zoning of Soil Erosion Control

6.1 General Requirements

6.1.1 The range of responsibility for soil erosion control of hydropower construction projects shall cover the permanent land requisition, temporary land occupation and other land areas used for the project, which shall be analyzed and determined considering the layout, construction planning, and land requisition and resettlement results of main works.

6.1.2 The zoning for soil erosion control shall facilitate the typical design of soil and water conservation measures, and shall coordinate with the design of main works.

6.2 Identification of Range of Responsibility for Soil and Water Conservation

6.2.1 If there are overlaps of reservoir inundation area with construction production and living sites, residues disposal areas, surface soil storage areas, quarries, borrow areas, and construction roads, the range of responsibility for soil erosion control shall not be counted twice.

6.2.2 The area of land requisition for centralized resettlements in the resettlement works, special item reconstruction works and newly reclaimed land works shall be included in the range of responsibility for soil erosion control.

6.3 Zoning of Soil Erosion Control

6.3.1 Zoning of soil erosion control shall be based on the results of field investigation and survey, and within its determined range, and consider the project layout, construction disturbance characteristics, construction sequence, landform characteristics, and soil erosion impacts.

6.3.2 The range of responsibility for soil erosion control of hydropower construction projects shall be divided into two first-level zones, namely, hydropower complex area and resettlement area, and should meet the following requirements:

1 The hydropower complex area should be divided into second-level zones, such as main works areas, residues disposal areas, temporary stockpile areas of excavated materials, surface soil storage areas, road areas, quarries, borrow areas, construction production and living sites, and reservoir inundation areas.

2   The resettlement area should be divided into second-level zones, such as rural centralized resettlement areas, market town relocation areas, special item reconstruction areas.

7 Prediction and Analysis of Soil Erosion and Water Loss

7.1 General Requirements

7.1.1 The prediction of soil erosion and water loss shall include the prediction of soil loss amount and the analysis of soil erosion hazards.

7.1.2 For hydropower construction projects, the potential soil loss amount caused by project construction shall be predicted and the soil erosion and water loss hazards shall be analyzed for the most unfavorable situation in the absence of soil and water conservation measures, according to the local natural conditions and construction disturbance characteristics.

7.1.3 The prediction range of soil erosion and water loss shall be the range of responsibility for soil erosion control, and the prediction units shall be determined according to the similarity principles in terms of landform, disturbance mode, material composition of disturbed surface, and meteorological characteristics.

7.1.4 The prediction period of soil erosion and water loss shall cover the construction preparation period, construction period and natural recovery period, and the prediction period of each zone shall be determined according to the project construction schedule.

7.2 Basic Requirements of Each Stage

7.2.1 At the planning stage, the analysis and prediction of soil erosion and water loss shall use the qualitative analysis method, to preliminarily analyze the impacts and hazards of soil erosion caused by the plan implementation.

7.2.2 The analysis and prediction of soil erosion and water loss at the pre-feasibility study stage shall include the following:

1. The disturbed ground surface area shall be analyzed and determined on the basis of the preliminarily determined project area.

2. The residues disposal amount of the project shall be estimated according to the preliminary analysis result of the cut-fill balance of the main works.

3. The total and newly increased amounts of soil loss shall be preliminarily estimated, and the soil erosion modulus should be directly determined by empirical values or engineering analogy.

4. The potential soil erosion and water loss hazards caused by project

construction shall be preliminarily analyzed.

7.2.3 The analysis and prediction of soil erosion and water loss at the feasibility study stage shall include the following:

1 The links and factors that might cause soil erosion and water loss in project construction process shall be analyzed.

2 The disturbed ground surface and damaged vegetation area shall be counted and analyzed.

3 Based on the results of the construction planning, the total residues amount that might be generated from the project construction shall be predicted.

4 The prediction of soil loss amount shall mainly include the background value analysis of soil loss in the project area, the prediction of soil loss amount during project construction, and the calculation of newly increased soil loss amount. The prediction methods for soil loss of the project area and each prediction unit should comply with the current standards of China GB 50433, *Technical Standard of Soil and Water Conservation for Production and Construction Projects*; and SL 773, *Guidelines for Measurement and Estimation of Soil Erosion in Production and Construction Projects*.

5 The potential soil erosion and water loss hazards shall be analyzed.

7.3 Analysis of Prediction Results

7.3.1 The analysis of the prediction results at planning stage shall preliminarily propose the key areas for soil erosion control based on the qualitative analysis results of soil erosion.

7.3.2 The analysis of the prediction results at the pre-feasibility study stage shall propose the key areas and periods of soil erosion control based on the preliminary prediction results.

7.3.3 The analysis of the prediction results at the feasibility study stage shall mainly include the following:

1 Based on the prediction results of newly increased soil loss amount in each prediction unit, the prediction results of each zone shall be analyzed and summarized; the disturbed ground surface and damaged vegetation areas, the residues amount and the total amount of newly increased soil loss of the project shall be proposed; main links and periods of soil loss newly generated from the project shall be defined.

2   The key areas and periods that soil erosion and water loss hazards emerge shall be defined, and guidelines on the key areas for soil erosion control and the layout of soil and water conservation measures shall be proposed.

3   The key areas and periods of soil and water conservation monitoring during the project construction and operation shall be defined.

8 Goals and Measures Layout of Soil Erosion and Water Loss Control

8.1 General Requirements

8.1.1 The soil erosion control standards and goals shall be determined according to the sensitivity of soil and water conservation and the soil erosion impact degree of the project area, and shall comply with the current national standard GB/T 50434, *Standard of Soil Erosion Control for Production and Construction Projects*.

8.1.2 The general layout of soil erosion control measures shall set up protection against hazards, and adopt overall design and layout and scientific configuration based on the site conditions of the main works, the project general layout and design, natural conditions and the current status of soil erosion.

8.2 Goals and Standards of Soil Erosion and Water Loss Control

8.2.1 The goals of soil erosion and water loss control shall meet the following requirements:

1 The goals of soil erosion control shall be determined by periods.

2 In the range of responsibility for soil erosion and water loss control, the newly increased soil erosion shall be effectively restricted, and the original soil erosion and water loss shall be controlled.

3 The ecological environment of terrestrial vegetation in the project area shall be restored and improved. The vegetation measures standards shall consider the requirements of regional planning and operation management and, for the areas close to urban area, shall consider the requirements of urban planning.

8.2.2 The level of detail and tasks of each stage shall meet the following requirements:

1 The soil erosion and water loss control standards of the project shall be defined preliminarily at the pre-feasibility study stage.

2 The soil erosion and water loss control standards shall be defined at the feasibility study stage, and soil erosion control indicators shall be listed by periods of the project.

8.3 Layout of Soil Erosion and Water Loss Control Measures

8.3.1 The general layout of soil erosion control measures shall meet the following requirements:

1. The layout of soil erosion and water loss control measures shall take local experience in similar production and construction projects as reference.

2. Attention shall be paid to the protection of surface soil resources.

3. Attention shall be paid to the drainage, collection and utilization of precipitation and the connection between drainage and downstream, to protect the downstream from damage.

4. Attention shall be paid to the protection of residues disposal areas, quarries and borrow areas.

5. Attention shall be paid to surface protection by preventing ground surface exposure, giving priority to vegetation measures and limiting the hardened area.

6. Attention shall be paid to the temporary protection during the construction period. The temporary soil stockpiles and exposed surface shall be timely protected.

7. If there are needs for vegetation protection, native vegetation transplantation, and native plant seedling in the project area, measures layout should be combined with rare plants transplantation, native plant species heel-in, and native plant seedling.

8. The existing soil erosion and water loss within the range of responsibility shall be controlled.

8.3.2 The system of soil erosion and water loss control measures shall meet the following requirements:

1. The system of soil erosion control measures shall be presented in table and block diagram by zones, engineering measures, vegetation measures, and temporary protection measures.

2. The layout of soil erosion control measures shall consider the landform, geology, meteorology, hydrology, soil, and vegetation, and coordinate with the local socio-economy and ecological environment.

8.3.3 The layout of soil and water conservation measures in the hydropower complex area shall meet the following requirements:

1. The engineering measures shall coordinate with the design of the main works, and measures such as blocking, slope protection, interception and drainage, and land improvement shall be arranged.

2. The sections with vegetation measures shall be determined based

on the analysis of the general layout of the main works and the land occupation of buildings, roads and other facilities, and the layout of vegetation measures shall be determined according to the requirements of soil erosion control, site conditions and operation management of the sections. For the excavated slope section at the abutment, the vegetation measures shall be combined with engineering measures to achieve landscaping on the premise of satisfying the project safety.

3 The temporary measures in the hydropower complex area should be planned in a holistic manner by combining permanent measures with temporary measures where practical.

8.3.4 The layout of soil and water conservation measures in residues disposal areas shall meet the following requirements:

1 The recommended scheme for protection of residues disposal areas shall be proposed through comparison of alternatives in terms of the types, effects and costs of protection measures, taking into account the project safety, construction conditions, and material sources.

2 The measures of dregs blocking, flood prevention and drainage, slope protection and vegetation construction shall be taken, and a combination of engineering measures, vegetation measures and temporary measures shall be adopted, to form a complete system of protective measures and ensure the proper protection, safety and stability.

8.3.5 The temporary blocking, drainage, vegetation, covering and other measures shall be adopted in the surface soil storage area and the temporary stockpile area of excavated materials.

8.3.6 The layout of soil and water conservation measures in the quarries and borrow areas shall meet the following requirements:

1 Emphasis shall be placed on the protection for stripping off useless layers, surface soil and mining slopes.

2 Corresponding soil erosion control measures shall be arranged for waste materials during mining.

3 The land improvement, farmland restoration and vegetation restoration measures shall be arranged considering the original land use function or the planned land use.

8.3.7 The layout of soil and water conservation measures in the production and living sites shall meet the following requirements:

1 The temporary or permanent drainage measures shall be arranged around and within the sites according to the construction season, precipitation, land occupation and terrain; the temporary blocking or covering measures shall be arranged for the stockpile areas within the sites; temporary vegetation measures should be adopted for temporary living sites with a long construction period; the landscaping measures shall be adopted for the living site with both temporary and permanent purposes.

2 According to the land occupation types and the final land use of the production and living sites, the measures of land improvement, farmland restoration and vegetation restoration shall be adopted. For the sandstorm areas in northwestern China, necessary measures shall be adopted.

3 According to the requirements of operation management and landscaping, the layout of lawns, spray-seeding, ornamental arbors, shrubs and flowers planting, rainwater collection and utilization, and supporting irrigation shall be carried out considering the natural conditions of the project area.

8.3.8 The layout of soil and water conservation measures in the road area shall meet the following requirements:

1 Measures such as slope protection, blocking, interception and drainage, and vegetation restoration shall be taken for permanent roads.

2 The temporary drainage and blocking measures shall be arranged for temporary construction roads according to the terrain, precipitation, and impact on the surrounding areas. The measures such as land improvement, vegetation restoration or farmland restoration shall be arranged according to the planned land use.

3 The permanent drainage and vegetation measures should be arranged for the construction roads with both temporary and permanent purposes.

4 The protective measures such as gravel blanket, grass pane and sand fence should be adopted on both sides of construction roads with a long construction period in the sandstorm areas in northwestern China.

8.3.9 The soil and water conservation measures for resettlement areas shall be arranged according to resettlement planning.

9 Residues Disposal Area

9.0.1 The design of residues disposal area shall mainly include site selection and layout, stockpiling, stability analysis, and general layout of measures.

9.0.2 The site selection and layout of residues disposal area shall be determined by the topographical and geological conditions, layout of hydropower complex structures, construction site conditions, on-site roads, haul distance, rehandling, subsequent site utilization, taking into account the construction general layout.

9.0.3 Residues disposal areas must not be set in the areas which have significant impacts on important public facilities, infrastructures, industrial enterprises, residential areas, etc.

9.0.4 Residues disposal areas should not be set in the areas prone to debris flow; if inevitable, a special demonstration shall be made and corresponding control measures shall be taken to ensure safety.

9.0.5 Residues disposal areas should not be set within the management scope of river course; if inevitable, necessary analysis and demonstration shall be carried out to ensure the existing functions of the river, such as flooding, navigation, and water supply, and to avoid negative effects on the protected objects upstream and downstream. In addition, the setup of residues disposal areas shall acquire the consent of the river management department,

9.0.6 For the residues disposal area set in the reservoir-inundated area, priority should be given to the subsequent utilization of the top platform. For the residues disposal area set in the drawdown area, the impact of reservoir drawdown on the residues disposal area stability shall be fully considered.

9.0.7 The stockpiling of residues disposal area shall be designed as a whole and mainly include capacity, land occupation area, maximum stockpile height, the height and width of steps, width of stepped platform, natural angle of repose, and integrated slope.

9.0.8 The design of residues disposal area shall include stability analysis which shall be made according to the residues disposal area grade, topography and geology, taking into account stockpiling form, residues composition, physical and mechanical parameters, and corresponding working conditions. The stability evaluation of residues disposal area shall be carried out to analyze the physical and mechanical properties based on the survey in accordance with the status quo of residues storage, determine the working conditions and calculation method, carry out stability evaluation, considering the protective measures of residues disposal area, give conclusions of residues disposal area stability analysis, and propose the measures for safety and stability.

9.0.9 The system of soil and water conservation measures for a residues disposal area shall be selected and established according to its location, class, grade, topography, stockpile height, stability and surrounding safety, and subsequent utilization, taking into account residues composition, local climate, etc.

9.0.10 The design of residues disposal area shall comply with the current sector standards NB/T 35111, *Design Code for Spoil Areas of Hydropower Projects*; and NB/T 10344, *Code for Design of Soil and Water Conservation for Hydropower Projects*.

10 Protection and Utilization of Surface Soil Resources

10.0.1 The protection and utilization of surface soil resources shall mainly include stripping, stockpiling, protection, improvement, and comprehensive utilization of surface soil.

10.0.2 The range, thickness and quantity of surface soil stripping shall be determined according to field investigation and sampling, and the surface matured soil in the construction disturbance area should be selected. When the surface matured soil is insufficient, the non-matured soil in the project area should be used preferentially, and soil improvement experiments shall be conducted to explore improvement measures.

10.0.3 Surface soil stockpiling areas shall be sited by a comparison of alternatives, and should be arranged in the construction sites to reduce rehandling, disturbance and ground surface occupation, without affecting the construction.

10.0.4 Temporary measures such as blocking, drainage, covering, and vegetation shall be taken for surface soil protection after stockpiling.

10.0.5 The utilization of surface soil resources shall meet the following requirements:

1 Stripped surface soil shall be used for farmland restoration and vegetation restoration in the project area, or may be used for land improvement in other areas.

2 The subsequent utilization of remaining surface soil shall be specified, and the remaining surface soil shall be stored and temporarily protected.

10.0.6 The protection and utilization of surface soil resources shall consider the requirements of soil and water conservation, ecological restoration and reclamation in resettlement works.

11 Design of Soil Erosion and Water Loss Control Measures

11.1 Dregs-Blocking Works

11.1.1 The special stockyard and complete dregs-blocking works shall be set up for the residues generated from the project construction.

11.1.2 The types of dregs-blocking works shall mainly include dregs-blocking walls, dregs-blocking dikes, slag enclosing weirs and dregs-blocking dams, and the type shall be reasonably selected considering the type of residues disposal area, stockpiling scheme, topography, geology, and hydrology.

11.1.3 The design of dregs-blocking works shall follow the principle of reliability, economy and rationality.

11.1.4 The dregs-blocking works shall be coordinated with flood release and diversion works and land improvement works to meet the overall stability and safe operation requirements of residues disposal area.

11.2 Flood Release and Diversion Works

11.2.1 In the construction and operation of hydropower projects, flood release and diversion works shall be constructed according to the source and damage of the flood if the disturbance of the original landform leads to the change of runoff process and soil loss.

11.2.2 The types of flood release and diversion works shall include flood retaining works, flood release works, drainage works and debris flow control works, and the type shall be mainly determined by topography, hydrology and protected objects.

11.2.3 For the project with flood hazards from upstream small-basin gullies, priority should be given to flood release and diversion works which may be a combination of flood-control dam and flood release channel or culvert, and the corresponding design shall comply with the current sector standard NB/T 35121, *Code for Design of Gully Treatment for Hydropower Projects*.

11.2.4 In the case of large runoff from the surrounding slope surface and a high possibility of flood hazards, flood interception and drainage ditches and channels shall be constructed on the slope surface and at the slope toe according to the topographic and hydrological conditions, and coordinate with drainage works of the sites, roads and other ground surface within the project area to discharge flood safely.

11.2.5 Where there are debris flow gullies in and around the project area that might cause hazards, the flood release and drainage works shall be designed in combination with debris flow control.

11.2.6 The design of debris flow control works shall comply with the current sector standard DZ/T 0239, *Code for Design of Debris Flow Disaster Prevention Engineering*.

11.3 Slope Protection Works

11.3.1 Slope protection measures shall be taken for the stable slopes and residues slopes formed by excavation, filling, quarrying, and soil borrowing for main works.

11.3.2 Slope protection measures shall mainly cover slope toe protection, slope protection by engineering measures, vegetation measures or their combinations, slope drainage and seepage control measures, which shall be mainly determined by topography, meteorology, hydrology and geology, site conditions, and the importance of downstream protected objects.

11.4 Precipitation Storage and Infiltration Works

11.4.1 Precipitation storage and infiltration works shall mainly include rainfall collection, storage and utilization, reuse and infiltration works, which shall be determined by the project-specific conditions and the characteristics of the region where the project is located.

11.4.2 Rainfall collection, storage and utilization works shall adopt an overall layout of facilities for surface water collection, storage, conveyance and supply, which should be combined with the target measures corresponding to different utilization purposes.

11.4.3 Rainfall collection and reuse system shall include facilities for rainfall collection and storage, reuse water pipe network, etc.

11.5 Land Improvement Works

11.5.1 The land improvement works shall mainly cover the leveling and loosening of disturbed and occupied land, surface soil recovery, farmland leveling and ploughing, land improvement, as well as the restoration of necessary water system and water conservancy facilities.

11.5.2 The land improvement works shall be implemented upon the completion of the construction of each zone.

11.5.3 The land improvement works design shall be coordinated with the local economy and ecological environment, and shall take measures of water storage and soil conservation cultivation to facilitate vegetation restoration and land improvement, taking into account flood release and diversion works and vegetation works.

11.6 Windbreak and Sand Fixation Works

11.6.1 Windbreak and sand fixation works shall be arranged when the hydropower construction project is located in a windy-sandy region or the project-caused disturbance to landform or damage to vegetation induces land desertification.

11.6.2 Windbreak and sand fixation works shall mainly include engineering measures, vegetation measures and enclosure measures, which shall be determined according to the characteristics of local climate and sandstorm hazards in the project area. The design of windbreak and sand fixation works shall comply with the current national standard GB/T 16453.5, *Comprehensive Control of Soil and Water Conservation—Technical Specification—Technique for Erosion Control of Wind Erosion*.

Translator's Annotation: GB/T 16453.5 has replaced GB/T 16453 in this article.

11.7 Vegetation Construction Works

11.7.1 The hydropower project construction shall reduce the occupation of and damage to vegetation. Upon completion of project construction, site clearing and land improvement works shall be implemented to improve the site conditions and restore vegetation.

11.7.2 The configuration of vegetation construction works shall be comprehensively determined by the grade of vegetation construction works, the natural and social environment, climate conditions, site conditions, land use nature and vegetation requirements.

11.7.3 Vegetation construction works serves the purpose of soil erosion control, which shall take vegetation measures as primary considering the needs of ecological protection and landscaping in the project area, and may take engineering measures as supplementary if necessary.

11.7.4 The design objectives of vegetation construction works are to ensure the construction and operation of the project and improve the ecological environment of the project area and its surroundings.

11.7.5 The design of vegetation construction works shall specify the technical requirements for forest and grass planting, including land improvement, seedling and planting, tending, etc.

11.8 Temporary Protection Works

11.8.1 Temporary protection works are mainly applicable to the soil erosion control in the disturbed areas during construction of the project. Temporary protection works should be coordinated with permanent protection works.

11.8.2 Temporary protection works shall include the measures of blocking, slope protection, drainage, sediment deposition, covering, vegetation, etc.

11.8.3 Temporary protection works shall be adopted after analysis of the surface exposure time, region and rainfall conditions, taking into account the soil and water conservation analysis and assessment of main works and construction planning.

12 Construction Planning on Soil and Water Conservation

12.1 General Requirements

12.1.1 The construction planning on soil and water conservation shall be conducted reasonably according to the project-specific conditions, and shall make full use of the water, power and compressed air supply facilities, access roads, construction sites, and temporary facilities of main works.

12.1.2 The construction general layout of soil and water conservation works shall give consideration to the construction schedule and time sequence between the main works and the soil and water conservation works and between different unit works of soil and water conservation, and shall plan the construction sites rationally, coordinate the construction of different unit works, reduce the amount of excavation and waste disposal, and reduce rehandling.

12.1.3 For the soil and water conservation works near gullies or rivers, the flood impact on its construction shall be considered.

12.2 Technical Requirements

12.2.1 The construction conditions of soil and water conservation works shall meet the following requirements:

1. The permanent and temporary roads of main works shall be fully used and, when necessary, temporary construction roads may be provided.

2. The water and power supply modes of soil and water conservation works shall be determined, considering the water and power supply conditions of the main works and the surroundings. If the construction contractor is required to provide water and power supply, the design requirements shall be specified.

3. The sources of stone blocks, sand, cement, steel, wood, explosives, and fuel used for construction of soil and water conservation works should be consistent with those of the main works. The sources of seedlings and seeds shall be specified. The seedlings and seeds should be purchased locally and shall be provided with quarantine certificate. The corresponding design shall be proposed if there are special requirements for treatment and transportation of seedlings and seeds.

4. The surface soil stripped and stored nearby should preferably be used for vegetation.

12.2.2 The construction layout of soil and water conservation works shall meet the following requirements:

1 The construction layout of soil and water conservation works shall be coordinated with that of main works, and should make full use of the construction sites, warehouses and management rooms and other temporary facilities of main works but shall not affect the construction of main works.

2 The scope of land occupation for construction shall be planned reasonably to avoid areas with unfavorable geological conditions or good vegetation.

3 For the construction facilities near a river or gully, the flood impact during construction shall be considered, and the flood control standard should be a 5- to 10-year flood.

4 The aggregate processing system and concrete mixing system should make use of those of main works. Simple, small or mobile processing, production and mixing facilities should be selected if necessary.

12.2.3 The construction methods for soil and water conservation works shall meet the following requirements:

1 Slope excavation shall adopt top-down construction procedure to avoid secondary slope cutting. The high slopes requiring support shall be timely supported after excavation.

2 The earth-rock shall be backfilled in layers and compacted manually with frog tamper, and the same soil type should be used.

3 For the earthwork or residues disposal requiring compaction, machinery shall be used for compaction in layers.

4 The masonry shall be layered and staggered by the grouting method and cured in time.

5 The construction of reinforced concrete shall meet the engineering practice.

6 Mechanical operation should be adopted in land improvement works, and manual land preparation shall be supplemented if necessary.

7 For seedling, mellow soil and layering should be adopted, and the plants may be mounded by the excavated soil and shall be tended.

8 The seeding and turf laying should adopt manual operation. When the laying area is large, turf roll may be selected and laid mechanically. The grass seeds may be sowed after applying base fertilizer as needed.

12.2.4 The construction schedule of soil and water conservation works shall meet the following requirements:

1. The construction schedule of soil and water conservation works shall be coordinated with that of the main works and shall be prepared in the principle of "prevention and protection first". The construction duration and schedule shall be determined considering the work quantities of soil and water conservation works and shall be provided with a double-line Gantt chart.

2. The construction schedule of blocking dregs works shall be prepared following the principle of "block first and discard follows". Flood discharge, blocking, and interception and drainage measures shall be completed before the stockpiling of dregs; engineering measures such as slope protection, interception and drainage shall be implemented in time in other areas.

3. The construction of vegetation measures shall shorten the exposure time of the disturbed land. According to the growth characteristics of plants, the construction season shall be selected, and the time periods for land preparation, planting, seeding, and cutting shall be specified.

4. The construction schedule of temporary measures shall be arranged timely and effectively, and shall be implemented when conditions permit.

12.3 Implementation Requirements

12.3.1 At the detailed design stage, the design results shall include the soil and water conservation construction planning; the bidding documents shall contain the corresponding soil and water conservation requirements and the bid shall contain the corresponding construction instructions; the construction contractor shall submit the construction plan.

12.3.2 Before construction, the supervision contractor shall review the construction plan provided by the construction contractor. The construction contractor shall stick to the construction plan approved by the supervision contractor.

12.3.3 At the construction stage, the project owner shall urge the implementation of relevant construction requirements for soil and water conservation works.

13 Soil and Water Conservation Supervision

13.1 General Requirements

13.1.1 The soil and water conservation supervision for hydropower construction projects shall comply with the national laws, regulations and relevant rules. The supervision contractor shall perform the duties agreed in the supervision contract independently, honestly and scientifically, coordinate the relationships among the parties involved in the project, and handle the issues and disputes arising in the construction of the project justly and impartially.

13.1.2 The soil and water conservation supervision for hydropower construction project shall be conducted according to the requirements of each stage of the project.

13.2 Basic Requirements of Each Stage

13.2.1 At the pre-feasibility study stage of a hydropower construction project, the work requirements for soil and water conservation supervision shall be put forward, which shall mainly include the preliminarily determined scope, content, objectives and plan of supervision work.

13.2.2 The soil and water conservation supervision scheme for hydropower construction project shall be put forward at the feasibility study stage which shall meet the following requirements:

1. Specify the scope and focus of soil and water conservation supervision.

2. Specify the requirements for procedures, responsibilities, and regulations of soil and water conservation supervision.

3. Specify the main content and methods for controlling the quality, construction process, construction schedule, etc. of soil and water conservation works.

13.2.3 The outline, planning and implementation rules of soil and water conservation supervision shall be prepared and implemented at the construction stage of hydropower construction project, and shall meet the following requirements:

1. Specify the requirements for the procedure, methods, content, quality, schedule, cost control, etc. of the soil and water conservation works.

2. Outline the construction contract documents, design documents and drawings, supervision plan, construction planning and relevant technical materials.

3 Specify the applicable scope, items, specialties, staff and division of responsibilities of supervision.

4 Specify the approval procedures and application for commencement of works and section of works, and organize the acceptance.

5 Specify content, measures and methods for the quality, schedule and cost control.

6 The content, measures and methods of other construction safety and environmental protection, main content of contract and information management, and relevant supervision procedure and content of acceptance and handover of soil and water conservation works shall be included.

13.2.4 At the acceptance stages of hydropower construction project, the supervision contractor shall participate in the acceptance at river closure and impoundment stages and the special acceptance of soil and water conservation works, and submit the summary report of soil and water conservation supervision of each stage. The summary report of soil and water conservation supervision shall comply with the current sector standard NB/T 35119, *Acceptance Specification for Soil and Water Conservation Engineering of Hydropower Projects*.

13.3 Supervision Implementation

13.3.1 The implementation content and relevant procedures and requirements of the soil and water conservation supervision of hydropower construction project shall comply with the current sector standard NB/T 35063, *Code for Environmental Supervision of Hydropower Projects*.

13.3.2 According to the scale and characteristics of the project and surrounding environment of the project site, a supervision department shall be established for soil and water conservation supervision of hydropower construction project. The division of labor, objectives and responsibilities of supervisors shall be determined. The supervision regulations shall be formulated, mainly including the regulations on design documents and drawings review, design change management, commencement approval, quality accident handling, payment, signature, claim, reporting, interim and completion acceptance, etc.

13.3.3 Soil and water conservation supervision for hydropower construction project mainly include the supervision work at the stages of construction preparation, construction and acceptance, with an emphasis on the control of the quality, progress and cost of soil and water conservation works.

14 Soil and Water Conservation Monitoring

14.0.1 The monitoring design of soil and water conservation shall be conducted, and the monitoring shall be initiated prior to construction disturbance, and shall proceed in both construction period and operation period of the hydropower construction project.

14.0.2 The basis, scope, period, content, methods, frequency, monitoring points, working procedure and the manpower and material input for soil and water conservation monitoring shall be defined.

14.0.3 The monitoring scope of soil and water conservation shall be consistent with the range of responsibility for soil erosion control, covering all construction sites of the project. The monitoring scope can be divided into hydropower complex area and resettlement area.

14.0.4 The monitoring content of soil and water conservation shall mainly include the conditions of disturbed land, quarrying and soil borrowing, residues, soil erosion, and the implementation and effect of soil and water conservation measures.

14.0.5 Soil and water conservation monitoring shall be conducted by adopting multiple methods, including field investigation, ground observation, satellite remote sensing, unmanned aerial vehicle remote sensing, video surveillance, data access, mathematical models, etc.

14.0.6 The key objects of soil and water conservation monitoring shall be determined according to the results of soil erosion prediction and hazard analysis, which shall mainly include residues disposal areas, surface soil storage area, quarries, borrow areas, large excavation and filling sites, temporary stockpile areas of excavated materials, access road area, the downstream area of the flood release zone with large catchment, and surrounding area of important protected facilities.

14.0.7 The methods, frequency, corresponding results, etc. of the soil and water conservation monitoring shall comply with the current sector standard NB/T 10506, *Technical Specification for Soil and Water Conservation Monitoring of Hydropower Projects*.

15 Scientific Research on Soil and Water Conservation

15.1 General Requirements

15.1.1 Scientific research and tests shall be included in the scientific research work of soil and water conservation of hydropower construction projects.

15.1.2 For the hydropower construction projects involving sensitive areas, ecologically fragile areas or important ecological function areas, or with constraints on soil and water conservation, soil and water conservation scientific research shall be timely carried out after necessity analysis.

15.2 Scientific Research Requirements and Tasks of Each Stage

15.2.1 At the pre-feasibility study stage, scientific research projects and special demonstration directions for soil and water conservation shall be put forward according to the characteristics of main works and issues proposed at the planning stage.

15.2.2 The scientific research requirements and tasks at the feasibility study stage shall mainly include the following:

1 Propose the specific research content and the implementation plan for the soil and water conservation scientific research project.

2 The key scientific research directions should mainly include the mechanism and influencing factors of soil erosion, ecological restoration technology in the soil and water conservation for high and steep slopes, reservoir shores and ecological connected areas, the protection of virginal vegetation, and soil improvement technology.

15.2.3 The scientific research requirements and tasks at the construction stage shall mainly include the following:

1 Scientific research on soil and water conservation is promoted according to the project construction progress, and the scientific research and test program is adjusted and optimized in time.

2 The practices and design of soil and water conservation measures is adjusted and optimized timely according to the scientific research and test results.

3 The scientific research direction is adjusted or the scientific research content is added in time according to the issues arising in the project construction.

4 The scientific research results are applied to the soil and water conservation design and construction of the hydropower construction project in time.

15.3 Scientific Research Implementation

15.3.1 The scientific research implementation shall include the project establishment, outline preparation, research development, and the demonstration, acceptance and application of achievements.

15.3.2 The scientific research implementation shall be guaranteed in terms of regulations, organization, staffing and funding.

16 General Design of Soil and Water Conservation

16.1 General Requirements

16.1.1 The general design of soil and water conservation for hydropower construction project shall be carried out according to the soil and water conservation program and its approval opinions, the review opinions of the feasibility study report, the project lotting plan, and the scientific research results of soil and water conservation. The general design of soil and water conservation for hydropower construction project shall put forward the general design and "Three-Simultaneity" implementation plan of soil and water conservation works.

16.1.2 The general design and "Three-Simultaneity" implementation plan of soil and water conservation works for hydropower construction projects shall be carried out in a comprehensive manner for the soil and water conservation work during the implementation stage, and be implemented by periods, locations and responsible units.

16.2 Technical Requirements for General Design of Soil and Water Conservation

16.2.1 The general design of soil and water conservation shall meet the following main requirements:

1. The construction layout of residues disposal areas, quarries, borrow areas, access roads, etc., range of responsibility for soil erosion control, zoning for soil erosion control, and general layout of soil and water conservation measures shall be reviewed, and the reasons for their adjustments shall be explained.

2. Stripping, storage and utilization of surface soil shall be reviewed, and corresponding survey and special design work shall be carried out.

3. The volume of earth-rock and residues shall be reviewed, the layout, storage and protection plan of the residues disposal areas shall be improved, and corresponding survey and design work shall be carried out.

4. Scientific research, special demonstration content and implementation schedule shall be refined according to the actual situation of the project.

5. On the basis of itemization, the design of soil erosion measures shall be refined. The grades and design standards of soil and water conservation measures shall be reviewed, and the quantities of soil and water

conservation works shall be calculated.

6   The lotting plan of soil and water conservation works shall be specified.

7   The construction methods and quality requirements and construction planning shall be refined.

8   The management planning of soil and water conservation shall be refined, and the construction contractor of soil and water conservation works shall be specified.

9   Cost estimate and annual investment of soil and water conservation works shall be reviewed.

16.2.2 The general design principles of soil and water conservation shall meet the following requirements:

1   Vegetation measures of soil and water conservation shall focus on the of soil erosion control highlight biodiversity, and adopt a multi-forest and multi-grass configuration with native trees and grasses as the main species.

2   The quantities of engineering measures and vegetation measures shall be coordinated with those of resettlement works without repeat or omission.

16.3 Technical Requirements for "Three-Simultaneity" Implementation Plan of Soil and Water Conservation

16.3.1 The "Three-Simultaneity" implementation plan of soil and water conservation works mainly outline the content and requirements of management and control of design, construction, cost and acceptance related to "Three-Simultaneity".

16.3.2 The design management and control of soil and water conservation shall include the following:

1   Drawings delivery plan for soil and water conservation shall be proposed according to the design schedule of main works.

2   Arrangements of the soil and water conservation design for the next stage shall be put forward.

3   Special design requirements for soil and water conservation shall be put forward for important works.

16.3.3 The construction management and control of soil and water conservation shall meet the following requirements:

1 Construction plan, equipment and material supply plan, construction management measures, emergency plan, etc. shall be formulated for soil and water conservation measures for simultaneous construction with main works.

2 Relevant responsibilities and obligations of the parties involved in the development, design, construction, supervision, and monitoring during construction shall be defined, the management agency shall be specified and corresponding regulations on soil and water conservation shall be implemented.

3 The construction planning of soil and water conservation shall consider the actual situation, adapt to local conditions and make full use of the water supply, electricity supply, accesses, temporary facilities, etc. of the main works.

4 Flood control requirements shall be put forward for the engineering measures of soil and water conservation works prone to flood.

16.3.4 The cost management and control of soil and water conservation shall include the formulation and implementation of annual investment plans and lot investment plans for soil and water conservation works, fund management responsibilities and specific work of the parties involved in the development, design, construction, supervision, monitoring, etc.

16.3.5 The acceptance management and control of soil and water conservation shall meet the following requirements:

1 The soil and water conservation works shall be put into operation simultaneously with the main works. Before it is put into operation, the acceptance of the soil and water conservation works shall be carried out, including the acceptance at the stages of river closure and impoundment and the special acceptance of soil and water conservation works.

2 After the construction of important works is completed, the acceptance of unit works, sections of works and individual works shall be carried out.

3 An acceptance plan for soil and water conservation works shall be prepared, and the responsibilities and obligations of all parties involved shall be defined.

17 Cost and Benefit Analysis of Soil and Water Conservation

17.1 General Requirements

17.1.1 The cost of soil and water conservation shall include the special cost of soil and water conservation works and the cost of soil and water conservation function in the main works.

17.1.2 The principles, basis, methods, and price levels used in estimating the special cost of soil and water conservation works shall be consistent with those used for the main works, and comply with the current sector standard NB/T 35072, *Preparation Regulation for Special Investment on Soil and Water Conservation of Hydropower Projects*, as well as other relevant regulations, norms and standards. The costs for scientific research and special demonstration of water and soil conservation at each stage shall be calculated separately and tabulated at the feasibility study stage.

17.2 Cost of Soil and Water Conservation

17.2.1 The cost of soil and water conservation at each stage shall meet the following requirements:

1 At the planning stage, the cost of soil and water conservation should be roughly estimated based on the project characteristics and analogy with the existing hydropower construction projects in the region.

2 At the pre-feasibility study stage, the cost of soil and water conservation shall be estimated based on the quantities of soil and water conservation measures.

3 At the feasibility study stage, according to the quantities of soil and water conservation, the basic prices and unit prices of the project shall be analyzed and calculated, and the itemized cost, independent costs, total cost and annual investment of soil and water conservation shall be calculated.

4 At the detailed design stage, the budgeted cost shall be calculated according to the quantities at this stage and the budgeted unit price.

17.2.2 The special cost of soil and water conservation works involving different provinces (autonomous regions, municipalities) shall be listed by province and calculated separately.

17.3 Benefit Analysis of Soil and Water Conservation

17.3.1 The level of detail and tasks of each stage shall meet the following requirements:

1. At the pre-feasibility study stage, the benefit of soil and water conservation shall be analyzed qualitatively.

2. At the feasibility study stage, the benefit indicators of soil and water conservation shall be calculated by zones.

17.3.2 The benefit analysis shall take the controlled percentage of erosion area, proportion of soil erosion control, percentage of blocked dregs and soil, percentage of protected topsoil, percentage of recovered forestry and grass, percentage of the forestry and grass coverage as the target parameters for quantitative calculation using the statistical analysis method. The benefit analysis shall comply with the current national standard GB/T 50434, *Standard of Soil Erosion Control for Production and Construction Projects*.

17.3.3 The qualitative analysis of soil and water conservation benefits shall be conducted in terms of water resources, soil resources, ecological environment, socio-economic stability and development, taking into account the prediction of soil erosion hazards during project construction.

18 Management of Soil and Water Conservation Works

18.1 General Requirements

18.1.1 The management of soil and water conservation works shall include construction management and operation management. The soil and water conservation works management shall be undertaken by the project owner and carried out in the hydropower complex area and resettlement area respectively.

18.1.2 The requirements for the soil and water conservation management department and staffing shall be put forward based on the management organizational structure of the main works and the soil and water conservation management tasks.

18.1.3 The main responsibilities of the soil and water conservation management department or staff shall include: in the construction period, organizing, coordinating and supervising the construction of soil and water conservation works, and ensuring the orderly construction of soil and water conservation works and the implementation of soil erosion control measures; in the operation period, being responsible for the operation and maintenance of soil and water conservation works.

18.1.4 The management of soil and water conservation works shall advocate informatization of soil and water conservation to manage the soil and water conservation work performed by the parties involved in the development, design, construction, supervision, etc. in the construction and operation periods, using data collection and information integration. A soil and water conservation information platform should be established by hydropower station or river basin, depending on the actual needs.

18.2 Management in Construction Period

18.2.1 The management of soil and water conservation works in the hydropower complex area in the construction period shall meet the following requirements:

1 The project owner shall be in charge of the management, and the parties involved shall participate in the management in the whole process of construction.

2 The scope of management shall cover the management of design, construction, operation and maintenance of soil and water conservation works in the areas of permanent land requisition and temporary land occupation of the hydropower complex area.

3   The content of management shall include the setup of organizational structure; the management of design, construction, supervision, monitoring, inspection and acceptance of soil and water conservation works; the management of the source and use of funds.

4   The setup of organizational structure shall define the organizations and staff and their responsibilities of parties involved in soil and water conservation and related management departments.

5   Soil and water conservation management rules and regulations shall be formulated, the archives of soil and water conservation works shall be established, and construction information and soil and water conservation work shall be reported to competent authorities.

6   For the design management, the design of soil and water conservation works shall be carried out in each stage as required; scientific research and special demonstrations of soil and water conservation shall be carried out according to the actual needs of the project.

7   The project owner shall strengthen the organization and management of soil and water conservation, and strictly control the changes in soil and water conservation. If there is any change during the construction period, the project owner shall follow the change procedures required by the nation or sector.

8   A chapter or section on soil and water conservation shall be included in the bidding design of the main works. If possible, the soil and water conservation works shall be a separate lot. The construction content and requirements of soil and water conservation measures shall be clearly specified in the bidding documents and contract.

18.2.2 Construction management of soil and water conservation works shall meet the following requirements:

1   The construction contractor shall formulate a detailed construction schedule for the soil and water conservation program, strengthen the management of soil and water conservation works, to ensure the implementation of the "Three-Simultaneity" system in which the soil and water conservation works is designed, constructed, and put into operation simultaneously with the main works.

2   The construction contractor shall construct in strict accordance with the design drawings and the technical requirements, and meet the construction schedule.

18.2.3 The routine management and acceptance of soil and water conservation facilities shall meet the following requirements:

1 The soil and water conservation facilities in operation shall be inspected regularly or irregularly by dedicated persons, and problems detected shall be handled timely. Meanwhile, the soil erosion control in the project construction area and its impact on the surrounding area shall be checked. If there is any direct impact on the surrounding area, it shall be addressed in time and relevant requirements shall be put forward.

2 The acceptance of soil and water conservation works for hydropower construction projects shall be divided into stage acceptance and completion acceptance considering the acceptance procedures of hydropower construction projects and the characteristics of soil and water conservation works. The stage acceptance shall include the acceptance of soil and water conservation works at the stages of river closure and impoundment.

3 The acceptance of soil and water conservation works shall be completed before the stage acceptance and completion acceptance of the main works. The acceptance of soil and water conservation works may be divided into the acceptance of soil and water conservation works in hydropower complex area and resettlement area.

18.3 Management in Operation Period

18.3.1 The management of soil and water conservation works in the operation period shall mainly include the establishment of soil and water conservation management department and staffing, and the management of operational tasks, facilities and equipment, and costs.

18.3.2 The management of soil and water conservation works in the operation period shall meet the following requirements:

1 The plan for establishing the soil and water conservation management department and staffing shall be proposed according to the nature of the management department of the main works in the operation period. If there is a handover of the management departments from the construction period to the operation period, the responsibilities of the soil and water conservation management departments shall be defined.

2 The project owner shall be responsible for the management and maintenance of all soil and water conservation facilities built within the permanent land requisition of the hydropower construction project in

the operation period.

3   The safety operation and management requirements for the main structures and facilities of the soil and water conservation works shall be put forward.

4   For important residues disposal areas, important soil and water conservation facilities, or the areas with severe soil erosion due to reservoir-induced earthquake, landslides or other geological disasters, the requirements for regular monitoring during operation shall be put forward.

18.3.3 The facilities and equipment required for soil and water conservation management in the operation period shall be provided.

18.3.4 The sources of funds for soil and water conservation management in the operation period shall be defined, and the financial files shall be established accordingly.

Appendix A Contents of Soil and Water Conservation Chapter in the Hydropower Planning Report

1 General

1.1 Current Situation of Soil Erosion

1.2 Current Situation of Soil and Water Conservation

2 Analysis and Prediction of Soil Erosion Impact

2.1 Compliance Assessment of Relevant Planning of Soil and Water Conservation

2.2 Preliminary Analysis of Constraints of Main Works

2.3 Preliminary Analysis of Project Construction Impact on Soil Erosion

3 General Requirements and Preliminary Plan for Soil Erosion and Water Loss Control

3.1 General Layout and System of Soil and Water Conservation Measures

3.2 Preliminary Plan for Soil and Water Conservation Measures

3.3 Countermeasures for Constraints of Soil and Water Conservation

3.4 Rough Cost Estimate of Soil and Water Conservation

4 Conclusions and Suggestions on Soil and Water Conservation

4.1 Conclusions

4.2 Suggestions

NB/T 10509-2021

Appendix B Contents of Soil and Water Conservation Chapter in the Pre-feasibility Study Report of Hydropower Construction Project

1 General

2 Soil and Water Conservation Assessment

2.1 Preliminary Analysis and Assessment of Soil and Water Conservation

2.2 Conclusions and Suggestions

3 Range of Responsibility and Zoning for Soil Erosion Control

4 Prediction of Soil Erosion

4.1 Prediction of Soil Erosion

4.2 Prediction Conclusion and Comprehensive Analysis

5 Goals and General Layout of Soil Erosion Control Measures

5.1 Standards and Goals of Soil Erosion Control

5.2 General Layout of Soil Erosion Control Measures

6 Design of Soil and Water Conservation Measures by Zones

7 Monitoring and Management of Soil and Water Conservation

7.1 Soil and Water Conservation Monitoring

7.2 Soil and Water Conservation Management

8 Cost Estimate of Soil and Water Conservation

8.1 Basis

8.2 Methods

8.3 Cost Estimate

9 Conclusions and Suggestions on Soil and Water Conservation

9.1 Conclusions

9.2 Suggestions

Attached Drawings:

Figure 1 Range of Responsibility and General Layout of Measures for Soil Erosion Control

Figure 2 Typical Design Diagram of Soil and Water Conservation Measures

Attachments

Appendix C Contents of Soil and Water Conservation Chapter in the Feasibility Study Report of Hydropower Construction Project

1 General

1.1 Overview of Natural Environment

1.2 Current Situation of Soil Erosion

1.3 Current Situation of Soil and Water Conservation

1.4 Work Process of Soil and Water Conservation

1.5 Main Conclusions

2 Analysis and Assessment on Soil and Water Conservation

2.1 Analysis and Assessment on Project Site Selection

2.2 Analysis and Assessment on Construction Plan

2.3 Analysis and Assessment on Design of Main Works

2.4 Main Conclusions

3 Range of Responsibility and Zoning for Soil Erosion Control

3.1 Range of Responsibility for Soil Erosion Control

3.2 Zoning for Soil Erosion Control

4 Prediction of Soil Erosion

4.1 Prediction of Disturbed Surface Area

4.2 Prediction of Residues Volume

4.3 Prediction of Potential Soil Erosion

4.4 Prediction Conclusion and Comprehensive Analysis

5 Goals and General Layout of Soil Erosion Control Measures

5.1 Standards and Goals of Soil Erosion Control

5.2 General Layout of Soil Erosion Control Measures

6 Design of Soil and Water Conservation Measures

6.1 Design Basis of Soil and Water Conservation Measures

6.2 Zoning for Soil and Water Conservation and Design of Soil Erosion Control Measures

6.3	Quantities and Schedule of Soil and Water Conservation

7 Construction Planning of Soil and Water Conservation

8 Monitoring Plan of Soil and Water Conservation

9 Scientific Research and Special Demonstration Plan of Soil and Water Conservation

10 Special Cost on Soil and Water Conservation

10.1	Preparation Instructions
10.2	Basic Data
10.3	Cost Composition
10.4	Special Cost Estimation on Soil and Water Conservation

11 Soil and Water Conservation Management Plan

12 Conclusions and Suggestions

Attached Tables:

Table 1　Quantities and Implementation Schedule of Soil and Water Conservation Measures

Table 2　Summary Table and Attached Tables of Special Cost Calculation on Soil and Water Conservation

Attached drawings:

Figure 1　Range of Responsibility for Soil Erosion Control

Figure 2　Zoning for Soil Erosion Control and General Layout of Soil Erosion Control Measures

Figure 3　Layout of Residues Disposal Areas and Measures

Figure 4　Design Drawing of Soil and Water Conservation Measures

Figure 5　Layout of Soil and Water Conservation Monitoring Points

Attachments

Appendix D Contents of General Design and "Three-Simultaneity" Implementation Plan on Soil and Water Conservation of Hydropower Construction Project

1 General

1.1 Source of Task

1.2 Project Background

1.3 Design Purpose and Significance

1.4 Design Basis

1.5 Main Technical Standards

1.6 Level of Design of Main Soil and Water Conservation Measures

2 Project Overview

2.1 Project Development Tasks and Scale

2.2 General Layout of Hydropower Complex

2.3 Construction General Layout

2.4 Construction Schedule

2.5 Physical Progress

2.6 Project Lotting

3 Review of Previous Work and Overview of Current Work for Soil and Water Conservation

3.1 Review of Previous Work for Soil and Water Conservation

3.2 Implementation of Soil and Water Conservation Measures and Related Work at Current Stage

4 General Design of Soil and Water Conservation Measures

4.1 Design Principles

4.2 Design Tasks

4.3 Design of Soil and Water Conservation Measures

4.4 Scientific Research and Special Demonstration at Current Stage

5 Soil and Water Conservation Supervision and Monitoring

6   Soil and Water Conservation Management Plan

6.1   Soil and Water Conservation Management Goals

6.2   Soil and Water Conservation Management System

6.3   Management Organization and Responsibilities

6.4   Management Rules and Regulations

7   Special Cost of Soil and Water Conservation

7.1   Preparation Instructions

7.2   Cost Composition

7.3   Basic Prices

7.4   Cost Estimate

7.5   Cost Comparison with Feasible Study Stage

8   Lotting of Soil and Water Conservation Works

8.1   Lotting Principle

8.2   Lotting Items

8.3   Lotting Plan

8.4   Technology Roadmap for Soil and Water Conservation Works

9   Construction Planning of Soil and Water Conservation Works

10   "Three-Simultaneity" Implementation Plan for Soil and Water Conservation

10.1 Soil and Water Conservation Measures

10.2 "Three-Simultaneity" Implementation Plan and Schedule of Soil and Water Conservation Measures

10.3 Design Management and Control for Soil and Water Conservation

10.4 Construction Management and Control for Soil and Water Conservation

10.5 Cost Management and Control for Soil and Water Conservation

10.6 Acceptance Management and Control for Soil and Water Conservation

11   Conclusions and Suggestions

11.1 Conclusions

11.2 Suggestions

Attached Drawings:

Figure 1　Geographical Location Map

Figure 2　Construction General Layout

Figure 3　Construction Schedule

Figure 4　General Layout of Soil and Water Conservation Measures

Figure 5　Design Drawing of Soil and Water Conservation Measures by Zones

Figure 6　Layout of Soil Erosion Monitoring Points

Figure 7　Implementation Schedule of Soil and Water Conservation Measures

Attachments:

Attachment 1　Written Approval of Soil and Water Conservation Program

Attachment 2　Review Opinions on Feasibility Study Report

Attachment 3　Management Rules on Project Construction and Soil and Water Conservation

Appendix E Contents of Explanation of Construction Drawing Design of Soil and Water Conservation Works of Hydropower Construction Project

1 Project Overview

1.1 Basic Information

1.2 Overview on Design Work

2 Design Basis and Standards

2.1 Design Basis

2.2 Design Standards

3 Design Conditions and Basic Data

3.1 Design Conditions

3.2 Design Data

4 Drawing Description

5 Construction Technical Requirements and Instructions

Appendix F Contents of Summary Report on Soil and Water Conservation Design of Hydropower Construction Project

1　General

2　Project Overview

2.1　Design Basis

2.2　Project Overview

2.3　Project Area Overview

3　Project Design Overview

3.1　Design of Main Works

3.2　Construction General Layout

4　Soil and Water Conservation Design Process

4.1　Soil and Water Conservation Design at Feasibility Study Stage

4.2　Soil and Water Conservation Design at Detailed Design Stage

5　Soil and Water Conservation Design Outcomes and Changes

5.1　Design Outcomes

5.2　Design Changes

6　Conclusions and Suggestions

7　Attached Drawings and Attachments

7.1　Attached Drawings

7.2　Attachments

Appendix G Contents of Post Assessment Report on Soil and Water Conservation of Hydropower Construction Project

1   General

2   Project Overview

3   Soil and Water Conservation Goals

4   Implementation of Soil and Water Conservation

4.1   Preparation of Soil and Water Conservation Program

4.2   Design of Soil and Water Conservation Measures

4.3   Design Changes of Soil and Water Conservation Works

4.4   Construction of Soil and Water Conservation Works

4.5   Acceptance of Soil and Water Conservation Works

5   Soil and Water Conservation Effects

6   Soil and Water Conservation Post-Assessment

6.1   Assessment on Goals

6.2   Assessment on Process

6.3   Assessment on Effects

7   Conclusions and Suggestions

7.1   Conclusions

7.2   Suggestions

8   Attached Drawings and Attachments

8.1   Attached Drawings

8.2   Attachments

Appendix H Contents of Soil and Water Conservation Program of Hydropower Construction Project

1 Executive Summary

1.1 Project Overview

1.2 Preparation Basis

1.3 Design Target Year

1.4 Range of Responsibility for Soil Erosion Control

1.5 Soil Erosion Control Goals

1.6 Assessment Conclusions on Soil and Water Conservation

1.7 Soil Erosion Predictions

1.8 Arrangement of Soil and Water Conservation Measures

1.9 Soil and Water Conservation Monitoring Scheme

1.10 Cost and Benefit Analysis of Soil and Water Conservation

1.11 Conclusions

2 Project Overview

2.1 Project Components and Layout

2.2 Construction Planning

2.3 Land Occupation

2.4 Cut and Fill Balance

2.5 Resettlement and Special Item Reconstruction and Relocation

2.6 Construction Schedule

2.7 Natural Conditions

3 Analysis and Assessment on Soil and Water Conservation

3.1 Assessment on Soil and Water Conservation for Site and Route Selection of Main Works

3.2 Assessment on Soil and Water Conservation Construction Plan and Layout

3.3 Definition of Soil and Water Conservation Measures in Main Works

Design

4 Analysis and Prediction of Soil Erosion

4.1 Current Situation of Soil Erosion

4.2 Analysis on Influencing Factors of Soil Erosion

4.3 Soil Erosion Prediction

4.4 Hazard Analysis of Soil Erosion

4.5 Guiding Opinions

5 Soil and Water Conservation Measures

5.1 Zoning of Soil Erosion Control

5.2 General Layout of Measures

5.3 Layout of Measures by Zones

5.4 Construction Requirements

6 Soil and Water Conservation Monitoring

6.1 Scope and Period

6.2 Content and Methods

6.3 Layout of Monitoring Points

6.4 Implementation Conditions and Results

7 Cost Estimate of Soil and Water Conservation

7.1 Cost Estimate

7.2 Benefit Analysis

8 Soil and Water Conservation Management

8.1 Organization Management

8.2 Follow-up Design

8.3 Soil and Water Conservation Monitoring

8.4 Soil and Water Conservation Supervision

8.5 Soil and Water Conservation Construction

8.6 Soil and Water Conservation Acceptance

Attached Tables:

Table 1　Range of Responsibility for Soil Erosion Control

Table 2　Soil and Water Conservation Goals

Table 3　Unit Price Analysis

Figures:

Figure 1　Geographic Location Map

Figure 2　Project Area Geomorphologic and Hydrographic Map

Figure 3　Project General Layout

Figure 4　Project Area Soil Erosion Intensity Map

Figure 5　Zoning of Prevention and Control Region of Soil Erosion and Water Loss

Figure 6　Range of Responsibility for Soil Erosion Control

Figure 7　Zoning of Soil Erosion Control and General Layout of Soil Erosion Control Measures

Figure 8　Design Drawing of Soil and Water Conservation Measures

Figure 9　Layout of Soil and Water Conservation Monitoring Points

Attachments:

Attachment 1　Hydropower Construction Project Planning Review and Approval Documents

Attachment 2　Other Relevant Review Opinions and Special Study Reports

Explanation of Wording in This Code

1. Words used for different degrees of strictness are explained as follows in order to mark the differences in implementing the requirements of this code:

    1) Words denoting a very strict or mandatory requirement:

        "Must" is used for affirmation; "must not" for negation.

    2) Words denoting a strict requirement under normal conditions:

        "Shall" is used for affirmation; "shall not" for negation.

    3) Words denoting a permission of a slight choice or an indication of the most suitable choice when conditions permit

        "Should" is used for affirmation; "should not" for negation.

    4) "May" is used to express the option available, sometimes with the conditional permit.

2. "Shall meet the requirements of…" or "shall comply with…" is used in this code to indicate that it is necessary to comply with the requirements stipulated in other relative standards and codes.

List of Quoted Standards

GB 50021,	*Code for Investigation of Geotechnical Engineering*
GB 50433,	*Technical Standard of Soil and Water Conservation for Production and Construction Projects*
GB/T 50434,	*Standard of Soil Erosion Control for Production and Construction Projects*
GB 50487,	*Code for Engineering Geological Investigation of Water Resources and Hydropower*
GB/T 51297,	*Standard for Investigation and Survey of Soil and Water Conservation Projects*
GB/T 16453.5,	*Comprehensive Control of Soil and Water Conservation—Technical Specification—Technique for Erosion Control of Wind Erosion*
NB/T 10344,	*Code for Design of Soil and Water Conservation for Hydropower Projects*
NB/T 10506,	*Technical Specification for Soil and Water Conservation Monitoring of Hydropower Projects*
NB/T 10510,	*Technical Code for Eco-Restoration of Soil and Water Conservation for Hydropower Projects*
NB/T 10512,	*Code for Slope Design of Hydropower Projects*
NB/T 35046,	*Code for Calculating Design Flood of Hydropower Projects*
NB/T 35063,	*Code for Environmental Supervision of Hydropower Projects*
NB/T 35072,	*Preparation Regulation for Special Investment on Soil and Water Conservation of Hydropower Projects*
NB/T 35095,	*Code for Hydrologic Calculation of Small Watershed of Hydropower Projects*
NB/T 35111,	*Design Code for Spoil Areas of Hydropower Projects*
NB/T 35119,	*Acceptance Specification for Soil and Water Conservation Engineering of Hydropower Projects*
NB/T 35121,	*Code for Design of Gully Treatment for Hydropower Projects*
DZ/T 0239,	*Code for Design of Debris Flow Disaster Prevention*

SL 773, *Engineering Guidelines for Measurement and Estimation of Soil Erosion in Production and Construction Projects*